Section 4 Mechanics and materials

Topic 1 Force, energy and momentum

Scalars and vectors

Scalar and vector quantities

Physical quantities can be classified into two distinct categories:

- **scalar** quantities, which have only magnitude (size)
- **vector** quantities, which have both magnitude and direction

Table 1

Some common scalar quantities related to motion	Some common vector quantities related to motion
Mass	Weight
Speed	Velocity
Distance	Displacement

1 Which of the following quantities represents a vector quantity? (AO1   **1 mark**

A   the volume of a book

B   the distance between a book and the edge of a desk

C   the weight of a book acting on a desk

D   the mass of a book

2 The M25 orbital motorway is 188 km long. Late at night, a van takes 2 hours 15 minutes to complete one orbit of the motorway. Calculate:

a   the average speed of the van in km h^{-1} (AO2)   **1 mark**

b   the average velocity of the van in km h^{-1} (AO2)   **1 mark**

Addition of scalars and vectors

Scalar quantities can be added or subtracted as numbers. When vector quantities are added, their direction needs to be considered as well as their size. Vector quantities can be added together to determine the overall, **resultant** vector acting on an object. If two vectors are at right angles to each other, this is best done by using Pythagoras' theorem and trigonometry.

If two vectors are acting on an object at angles other than 90°, drawing a scale diagram, showing the size and direction of each vector, is a good way to find the sum of the two vectors. Vectors can be added in this way by drawing the 'tail' of one vector starting at the 'tip' of the other vector.

3 A sailing dinghy is blown across an estuary by a northerly wind exerting a force of 2500N. At the same time, a water current acts on the dinghy with a force of 1000N in a westerly direction. What is the resultant force acting on the dinghy? (AO1) `1 mark`

  A 2693N, 338°

  B 2693N, 22°

  C 3500N, 338°

  D 3500N, 22°

4 Two plough-horses act at angles of 40° either side of the direction of motion of a plough. Both horses pull with a force of 4750N. Use a scale diagram to determine the resultant force on the plough. (AO2) `3 marks`

Resolution of vectors and forces in equilibrium

Any vector, such as a force, F, can be broken down or **resolved** into two perpendicular components. If the force acts at an angle θ to the horizontal, it can be resolved into a horizontal force, F_h, and a vertical force, F_v, where:

When forces are in **equilibrium**, their resultant force is zero. This means that the vector sum of the horizontal components of the forces is zero, as is the vector sum of the vertical components.

$$F_h = F\cos\theta \quad \text{and} \quad F_v = F\sin\theta$$

5 A pendulum hangs at an angle of 15° to the vertical and is held stationary by a horizontal force of 3.2 N. What is the weight of the pendulum bob? (AO2)   **1 mark**

A   0.86 N

B   11.9 N

C   12.4 N

D   0.83 N

6 A car of mass 950 kg, initially at rest, starts to roll down a hill with a slope of 30° to the horizontal. The force of friction acting on the car as it rolls is 750 N parallel to the slope. Calculate the acceleration of the car. (AO2)   **4 marks**

Moments

Moments and couple

A turning force applied about a pivot point is called a **moment**. The size of a moment is determined by the size of the force and the perpendicular distance from the pivot point to the line of action of the force, and is defined as:

moment = force applied × perpendicular distance from the pivot

A pair of equal and opposite coplanar (parallel) forces acting about a shared pivot point, such as the forces used to turn a steering wheel, is called a **couple**, and is defined as:

moment of couple = force applied × perpendicular distance between the lines of action of the forces

7 A force of 100 N is applied at an angle of 45° to the handle of a wrench, as shown in Figure 1.

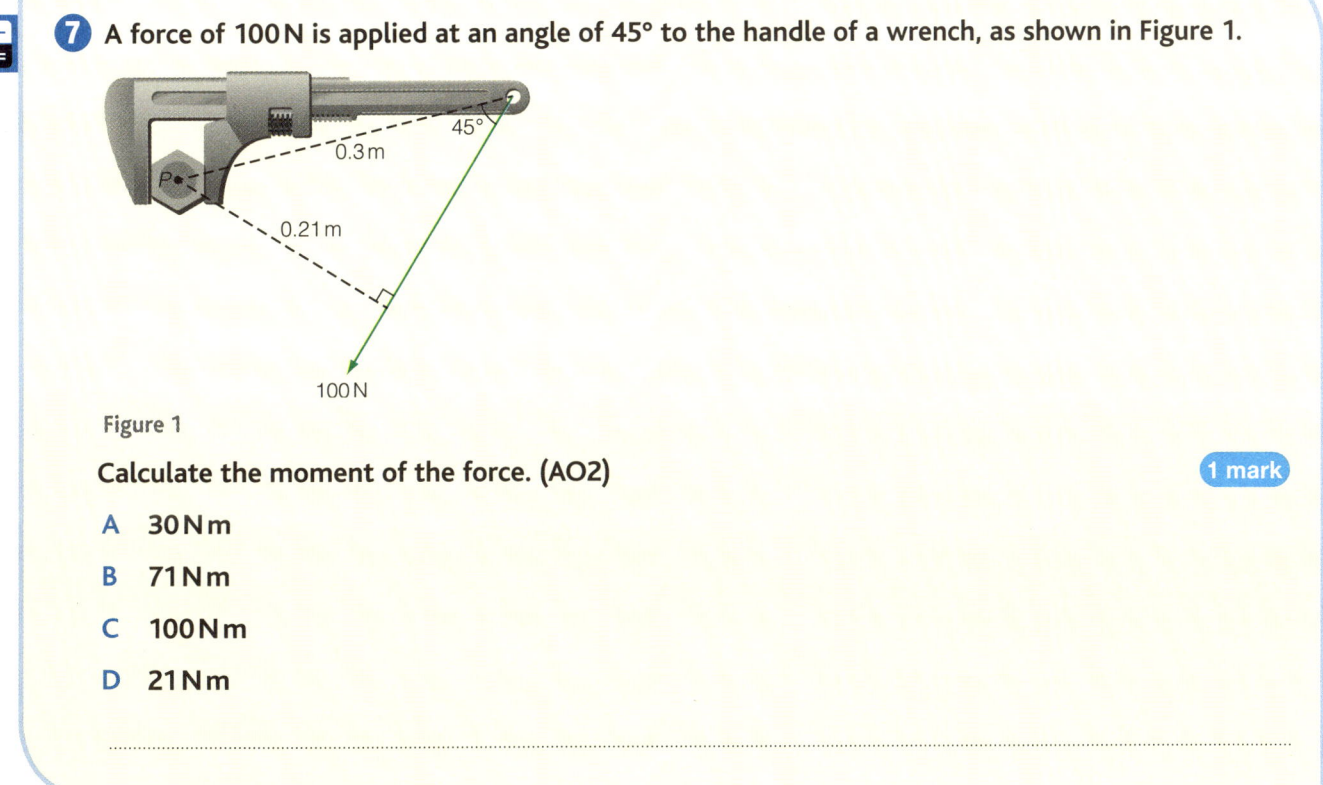

Figure 1

Calculate the moment of the force. (AO2)   **1 mark**

A   30 N m

B   71 N m

C   100 N m

D   21 N m

Principle of moments

An object at rest is in equilibrium and Newton's first law tells us that the forces acting on the object must be balanced. The moments acting on any object in equilibrium (about any pivot) must also be balanced, and hence the resultant moment acting about a pivot must be zero. This is called the **principle of moments** and, as moments can act in a clockwise or anticlockwise direction, is defined as:

sum of the clockwise moments
= sum of the anticlockwise moments

8 A boy and a girl balance their mother sitting on the opposite side of a see-saw. The mother has a weight of 590 N and sits 1.6 m from the pivot of the see-saw. The boy has a weight of 280 N and sits 0.80 m away from the pivot. The girl has a mass of 320 N. How far away from the pivot does the girl sit? (AO2)

1 mark

A 2.25 m

B 2.50 m

C 2.00 m

D 1.75 m

9

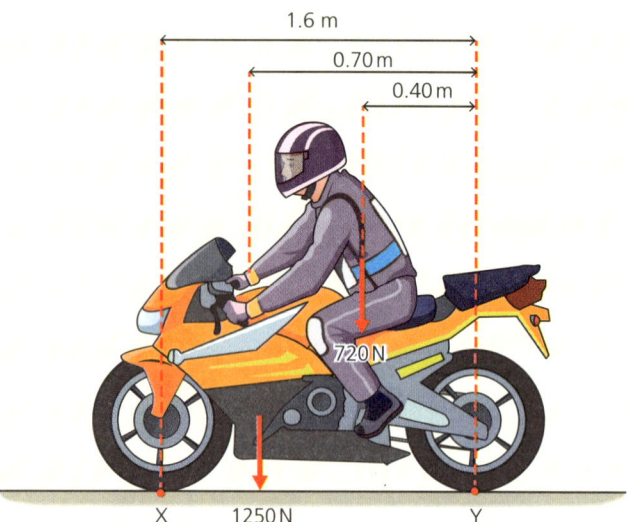

1.6 m
0.70 m
0.40 m
720 N
X   1250 N   Y

Figure 2

a Using the principle of moments, calculate the force of the front tyre on the road at X in Figure 2. (AO2)

2 marks

b Calculate the force of the back tyre on the road at Y. (AO2)

2 marks

Centre of mass

The **centre of mass** of an object is the single point in an object around which the resultant moment due to the pull of gravity is zero. The effect of this is that if an object is pivoted directly below its centre of mass, the object will balance and not topple over. For a uniform regular solid, the centre of mass is at its centre.

The centre of mass of an object is useful in mechanics as it enables extended objects to be considered as single points, thus simplifying the physics.

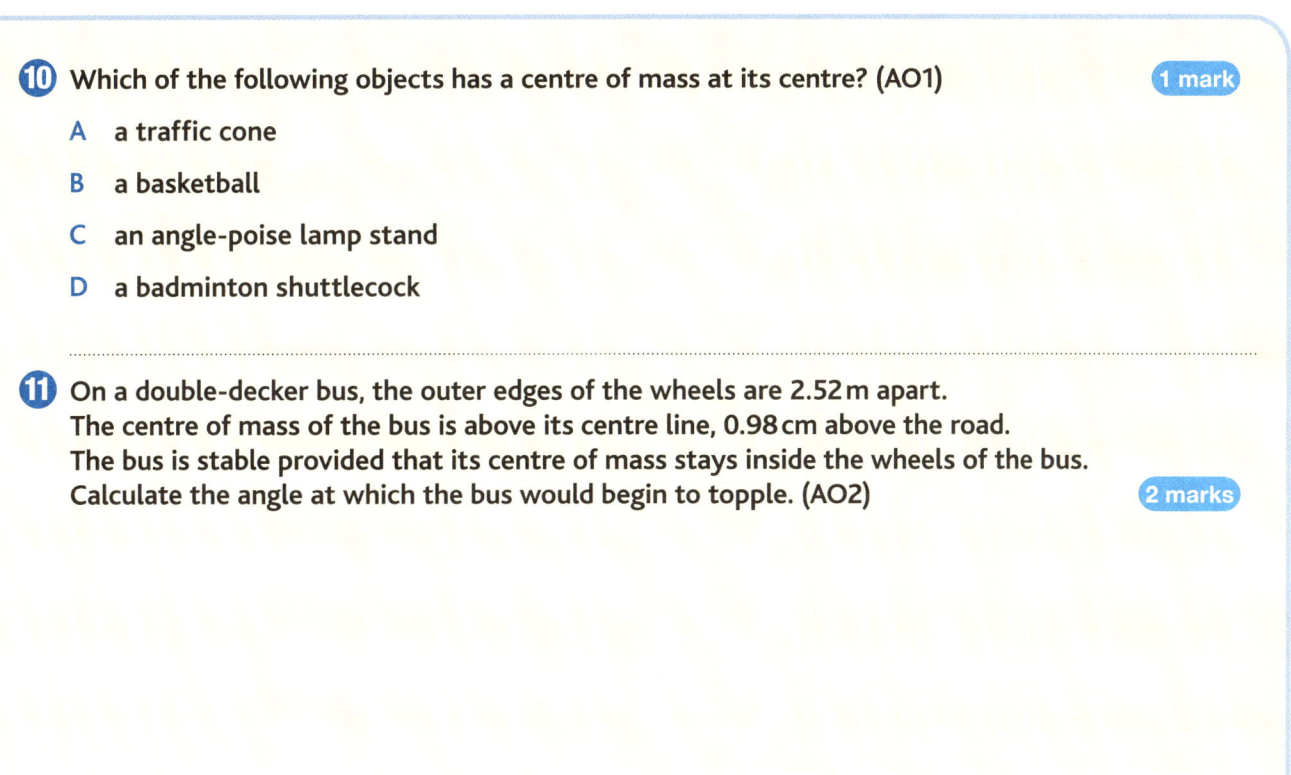

10 Which of the following objects has a centre of mass at its centre? (AO1)   `1 mark`

    A   a traffic cone

    B   a basketball

    C   an angle-poise lamp stand

    D   a badminton shuttlecock

11 On a double-decker bus, the outer edges of the wheels are 2.52 m apart. The centre of mass of the bus is above its centre line, 0.98 cm above the road. The bus is stable provided that its centre of mass stays inside the wheels of the bus. Calculate the angle at which the bus would begin to topple. (AO2)   `2 marks`

Motion along a straight line

Describing motion

The **displacement** of an object, s, is a vector quantity that describes the distance of the object from an origin, in a given direction. The (average) **speed** of an object is a scalar measurement of the distance that the object moves in a set time. The **velocity** of an object, v, is the vector speed of the object in a given direction:

$$v = \frac{\Delta s}{\Delta t}$$

The rate of change of velocity of an object is called its **acceleration**, a. Acceleration is also a vector quantity and is always defined with a direction, usually positive in one direction and negative in the opposite direction. Acceleration is measured in m s⁻² and is given by:

$$a = \frac{\Delta v}{\Delta t}$$

Uniform motion involves objects moving at constant speed (or being stationary), and non-uniform motion involves objects that are accelerating.

Speed and velocity can be given as an average or instantaneous value. If an object has an initial velocity, u, and a final velocity, v, the average velocity (or speed, if direction is not important) is given by:

$$\text{average velocity} = \frac{u + v}{2}$$

The motion of an object is best displayed graphically. There are two main ways of doing this: a displacement–time graph and a velocity–time graph. The gradient of a displacement–time graph equals the velocity of the object:

$$v = \frac{\Delta s}{\Delta t}$$

An object that is accelerating has a curved displacement–time graph. Velocity–time graphs are often much more useful than displacement–time graphs. The area under a velocity–time graph is equal to the distance travelled by the object in the given time interval, and the gradient of the graph at any point equals the acceleration of the object at that time.

12 Which graph in Figure 3 best represents the velocity–time graph for a helicopter landing? (AO3)

1 mark

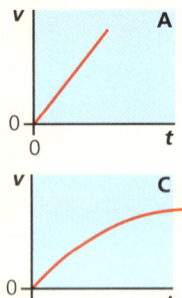 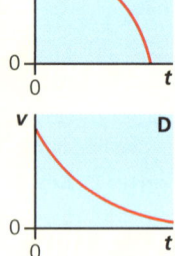

Figure 3

---

13 Figure 4 shows the motion of a car between two sets of traffic lights (at A and D) on a straight road.

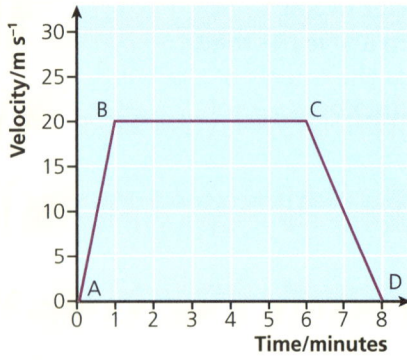

Figure 4 The velocity–time graph for a car travelling between two sets of traffic lights

Calculate the following quantities:

a the acceleration of the car from A to B (AO2)

1 mark

b the acceleration of the car from C to D (AO2)

1 mark

c the total distance travelled by the car between the two sets of traffic lights (AO2)

2 marks

Equations for uniform acceleration

The motion of an object with uniform acceleration can be described by four equations of motion, defined using the (suvat) symbols and equations shown in Table 2.

Table 2

Symbols	Equations
s displacement (m)	$v = u + at$
u initial velocity of the object ($m\,s^{-1}$)	$s = \left(\dfrac{u+v}{2}\right)t$
v final velocity of the object ($m\,s^{-1}$)	
a acceleration ($m\,s^{-2}$)	$s = ut + \dfrac{at^2}{2}$
t time (s)	$v^2 = u^2 + 2as$

14 A train undergoes uniform acceleration from rest to a constant velocity of $35\,m\,s^{-1}$ in a time of 1 minute and 10 s. What is the total distance travelled by the train during this time? (AO2)   **1 mark**

- A   1500 m
- B   3000 m
- C   1225 m
- D   19 m

15 A car undergoes uniform acceleration from rest onto a motorway slip road and joins the main carriageway 8.0 s later travelling at $31\,m\,s^{-1}$. The car then travels at this velocity for 20 s before accelerating at $2.2\,m\,s^{-2}$ for 5.0 s to overtake a lorry. Calculate:

- a   the acceleration of the car on the slip road (AO2)   **2 marks**

- b.   the distance travelled by the car on the slip road (AO2)   **1 mark**

- c   the final velocity of the car (AO2)   **1 mark**

- d   the total distance travelled by the car from rest to just overtaking the lorry (AO2)   **3 marks**

Acceleration due to gravity

Close to the surface of the Earth, objects moving under the influence of gravity accelerate towards the Earth at a constant rate of $g = 9.81\,m\,s^{-2}$. The value of g can be determined experimentally by timing an object (a ball) falling between two light gates, A and B. The light gates record the time at which the ball falls through each beam. Knowing the diameter of the ball, the velocity at each light gate can be calculated. The light gates also measure the time taken for the ball to travel between the two gates, t, hence:

$$g = \frac{v_B - v_A}{t}$$

9

16 In an experiment to measure g, a ball of diameter 1.32 cm falls through a top light gate in 8.30 ms and through a bottom light gate in 2.80 ms. The time taken to fall between the two light gates is 318 ms. The acceleration due to gravity is calculated to be: (AO2) `1 mark`

    A $9.82\,\text{m s}^{-2}$

    B $9.81\,\text{m s}^{-2}$

    C $9.80\,\text{m s}^{-2}$

    D $9.79\,\text{m s}^{-2}$

17 In a tip-off in a basketball game, the referee throws the basketball vertically upwards with an initial velocity of $11.3\,\text{m s}^{-1}$. The referee releases the ball at a height of 2.1 m. Assuming that air resistance is negligible, calculate the maximum height of the ball above the court. (AO2) `4 marks`

Projectile motion

Falling sideways

When an object moves sideways in a uniform gravitational field, the effect of gravity is to pull the object downwards at the same time as it moves sideways. The object will travel in a parabolic arc with a horizontal component of velocity, v_h, and a vertical component of velocity, v_v. These two components act independently of each other.

18 A golfer hits a drive off the tee at an angle of 25° to the horizontal at $90\,\text{m s}^{-1}$. Calculate the maximum height of the flight of the ball. (AO2) `1 mark`

    A   39 m

    B   92 m

    C   74 m

    D   26 m

19 A rugby player kicks a ball from the ground with a velocity of $20\,\text{m s}^{-1}$ at an angle of 35° to the horizontal. Calculate:

    a   the vertical and horizontal components of the initial velocity of the ball (AO2) `2 marks`

b the total flight time of the ball until it hits the ground again (AO2)   2 marks

....

....

....

....

....

c the horizontal distance travelled by the ball before it hits the ground (AO2)   1 mark

....

....

Moving through air

When an object moves through a fluid, a frictional force acts against the motion of the object. This is called a **drag** force and depends upon: the speed of the object, its cross-sectional area and the density of the fluid. An object falling through air, such as a parachutist, experiences a drag force upwards due to the air resistance as they fall downwards. This drag force increases with the speed of the parachutist and their cross-sectional area until eventually, in freefall, the size of the drag force equals the weight of the parachutist. This produces a zero resultant force and the parachutist stops accelerating and travels at a constant **terminal speed**.

20 The drag force on a golf ball in flight does not depend upon: (AO1)   1 mark

    A the angle of the flight of the ball to the horizontal

    B the cross-sectional area of the ball

    C the vertical component of the velocity of the ball

    D the horizontal component of the velocity of the ball

....

21 The weight of a parachutist is 700 N. Describe the forces acting on the parachutist and her motion as she falls from the point of release to when she begins to travel at terminal speed. (AO3)   6 marks

....

....

....

....

....

....

....

....

....

Newton's laws of motion

Newton's three laws of motion are:

First law: an object will remain at rest or continue to move in a straight line with a constant velocity unless it is acted on by an unbalanced force. This helps us to examine the forces acting on objects when the resultant force is zero using **free-body force diagrams** which show only the forces acting on a single body or object.

Second law: when an unbalanced (or resultant) force acts on an object, it will accelerate in the direction of the resultant force. The size of the acceleration is determined using the equation:

$F = ma$

Newton's second law of motion, and the equation, dictate what happens to an object when the resultant force is not zero.

Third law: when object A exerts a force, *F*, on object B, object B exerts an equal and opposite force, *F*, on object A. The pair of forces has the following properties:

- They have the same magnitude (size).
- They act in opposite directions.
- They act along the same line.
- They act on two separate bodies.
- They are always the same type of force, for example two contact forces or two gravitational forces.

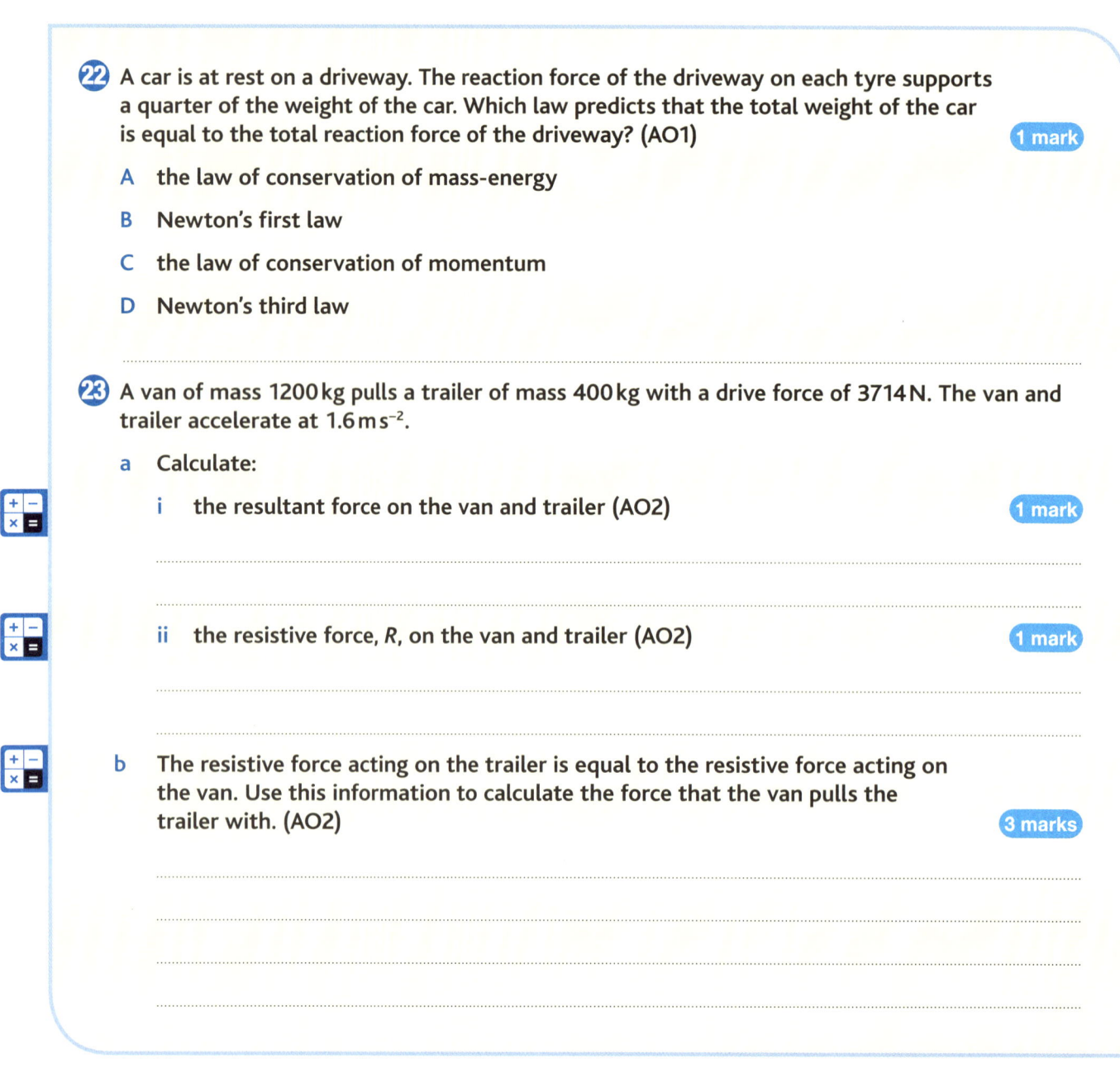

22 A car is at rest on a driveway. The reaction force of the driveway on each tyre supports a quarter of the weight of the car. Which law predicts that the total weight of the car is equal to the total reaction force of the driveway? (AO1) `1 mark`

   A   the law of conservation of mass-energy

   B   Newton's first law

   C   the law of conservation of momentum

   D   Newton's third law

23 A van of mass 1200 kg pulls a trailer of mass 400 kg with a drive force of 3714 N. The van and trailer accelerate at 1.6 m s⁻².

   a   Calculate:

      i   the resultant force on the van and trailer (AO2) `1 mark`

      ii   the resistive force, *R*, on the van and trailer (AO2) `1 mark`

   b   The resistive force acting on the trailer is equal to the resistive force acting on the van. Use this information to calculate the force that the van pulls the trailer with. (AO2) `3 marks`

Momentum

Momentum and conservation of momentum

The principle of **conservation of momentum** during interactions between objects allows us to predict the behaviour of the objects after the interactions. The momentum of an object is defined as:

momentum = mass × velocity   or   momentum = mv

Momentum is a vector quantity so can be represented by an arrow on scale diagrams. It has the unit kg m s^{-1}. Assuming that there are no external forces acting, momentum is conserved (the quantity remains constant) in all interactions between objects, therefore:

sum of momentums before an interaction
= sum of momentums after an interaction

There are two types of collisions (or explosions) between objects:
- **elastic collisions**, where kinetic energy is conserved as well as momentum
- **inelastic collisions**, where kinetic energy is not conserved, it is transformed into other forms of energy, such as sound, heat and strain energy. Momentum is still conserved during inelastic collisions.

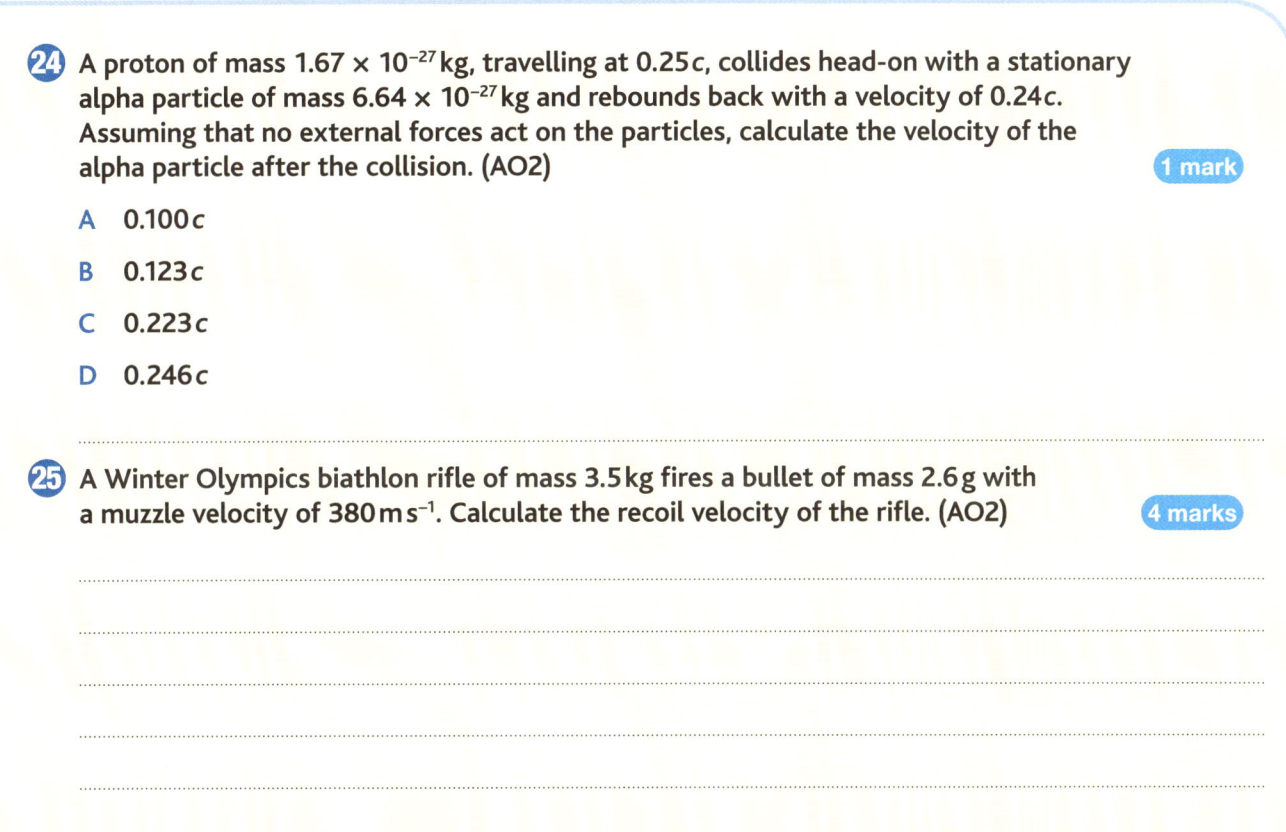

24 A proton of mass $1.67 \times 10^{-27}\,\text{kg}$, travelling at $0.25c$, collides head-on with a stationary alpha particle of mass $6.64 \times 10^{-27}\,\text{kg}$ and rebounds back with a velocity of $0.24c$. Assuming that no external forces act on the particles, calculate the velocity of the alpha particle after the collision. (AO2) **1 mark**

A $0.100c$

B $0.123c$

C $0.223c$

D $0.246c$

25 A Winter Olympics biathlon rifle of mass $3.5\,\text{kg}$ fires a bullet of mass $2.6\,\text{g}$ with a muzzle velocity of $380\,\text{m s}^{-1}$. Calculate the recoil velocity of the rifle. (AO2) **4 marks**

Force and impulse

Newton's second law, $F = ma$, can be rewritten as:

$$F = ma = m\frac{\Delta v}{\Delta t} = \frac{\Delta(mv)}{\Delta t}$$

In other words, **force** is the rate of change of momentum of an object. This equation also allows us to define another quantity called **impulse**, the change in momentum:

$$\text{impulse} = F\Delta t = \Delta(mv) = mv_{\text{final}} - mv_{\text{initial}}$$

The unit of impulse is Ns. If the force acting on an object is not constant, the impulse applied to it can be determined by measuring the area under the force–time graph for the object.

26 A cricket ball of mass 163 g, travelling at 28 m s⁻¹, is hit back by a batsman at a velocity of 36 m s⁻¹. Calculate the impulse that the bat exerts on the ball. (AO2) 1 mark

A   10 000 kg m s⁻¹

B   1300 kg m s⁻¹

C   10 kg m s⁻¹

D   1.3 kg m s⁻¹

27 Figure 5 shows the force exerted by a seatbelt on a passenger in a car during an emergency stop.

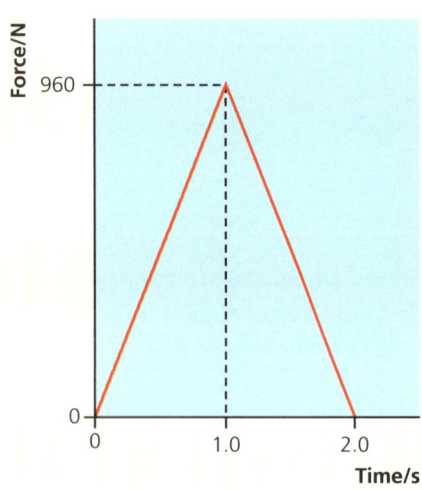

Figure 5

a   Use the graph to determine the change of momentum of the passenger. (AO3) **2 marks**

b   The seatbelt is replaced with a new one that produces a total impact time of 3.0 s. Calculate the maximum force exerted by the seatbelt on the passenger in a similar emergency stop. (AO2) **2 marks**

c   Explain why seatbelts in a car should be changed following their use in a collision. (AO1) **3 marks**

Work, energy and power

Work and energy

Work, W, is done when a force, F, moves an object in the direction of the force a distance, s, where:

$W = Fs$

Work is an energy change so is measured in joules, J. Forces acting at right angles to the direction of motion do no work. If the force acts at an angle, θ, to the direction of motion, the component of the force in the direction of motion is used, so that:

$W = Fs\cos\theta$

In some situations, the force acting on a moving object is not constant. The force required to stretch a spring or rubber band is a good example of this. To calculate the work done in stretching a spring, we need to consider the work done, ΔW, in stretching the spring through a small distance, Δs:

$\Delta W = F \times \Delta s$

The total work done in stretching the spring by an extension, s, in the direction of the force equals the area under its force–extension graph, between the origin and s.

Power and efficiency

Power, P, is the rate of doing work, or the rate of energy transfer, and is measured in watts, W, where:

$$P = \frac{\text{work done}}{\text{time taken}}$$

$$P = \frac{\Delta w}{\Delta T}$$

The power needed to keep an object, such as a vehicle, moving at a constant speed, v, can be calculated by substituting for the work done:

$$P = \frac{F \times s}{t}$$

$$P = Fv$$

The **efficiency** of an energy transfer can be expressed as a fraction or a percentage. Efficiency measures the amount of useful power output by a system compared to the total input power:

$$\text{efficiency} = \frac{\text{useful output power}}{\text{total input power}} \ (\times \ 100\%)$$

---

28 Which of the following is a unit of power? (AO1)   `1 mark`

A $Nm^{-1}s$

B Nms

C $Nm^{-1}s^{-1}$

D Nms^{-1}

---

29 Two men on either side of a canal take 12 minutes to pull a barge along the canal between two locks 200 m apart. Each man exerts a force of 100 N on the barge, both at an angle of 60° to the direction of motion of the barge.

a   Calculate:

i   the total force on the barge in the direction of motion (AO2)   `1 mark`

ii the total work done by the men pulling the barge (AO2) **1 mark**

iii the power exerted by each man (AO2) **2 marks**

b Explain why the total work done by the men's muscles will be much greater than the total work done on the barge. (AO1) **2 marks**

Conservation of energy

Any moving object possesses kinetic energy, E_k, which depends upon the velocity of the object and its mass:

$$E_k = \frac{1}{2}mv^2$$

An object which is lifted off the ground against the pull of gravity is given gravitational potential energy, ΔE_p, which depends upon the mass of the object, the acceleration due to gravity, g, and the height difference, Δh, that the object is moved through:

$$\Delta E_p = mg\Delta h$$

If the object is then dropped so that it falls through the same height, Δh, if there is no friction or air resistance, all the gravitational potential energy will be transformed into an equal amount of kinetic energy E_k, hence:

$$\Delta E_p = E_k$$

$$mg\Delta h = \frac{1}{2}mv^2$$

The total energy of the dropped object is conserved throughout the fall. The **conservation of energy** during energy transfer is a fundamental general principle of physics. The total energy involved during any energy transfer remains constant. In other words, energy is neither created nor destroyed during an energy transfer.

---

30 During an Apollo 15 moon walk, astronaut David Scott simultaneously dropped a geology hammer and an eagle feather and they both hit the ground at the same time. Which of the following factors did not affect the velocity of the objects as they hit the ground? (AO1)　　**1 mark**

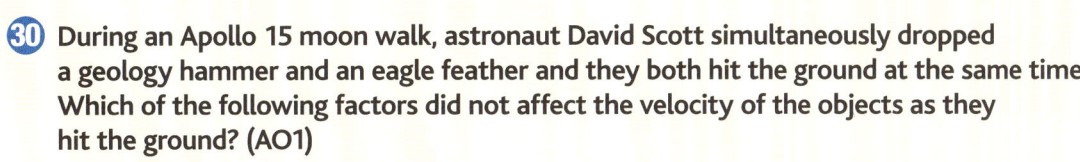

A　the mass of each object

B　the height that the objects were dropped from

C　the size of the acceleration due to gravity

D　the medium that the objects fell through

---

31 A rugby ball of mass 460 g is kicked vertically upwards by a player from ground level to a height of 30 m and then falls to the ground. Calculate:

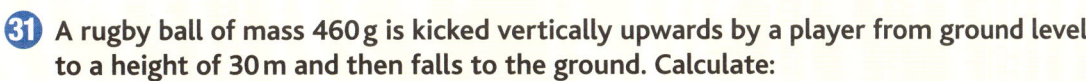

a　the gravitational potential energy of the ball at the top of its flight (AO2)　　**1 mark**

b　the velocity of the ball as it hits the ground, assuming that air resistance is negligible (AO2)　　**2 marks**

Exam-style questions

1 Coastal forts used to fire cannons at enemy ships which sailed too close to the shore. In one example of this, a cannon mounted in a fort 134 m above sea level hit an enemy ship at sea level, 300 m away from the base of the fort, as shown in Figure 6:

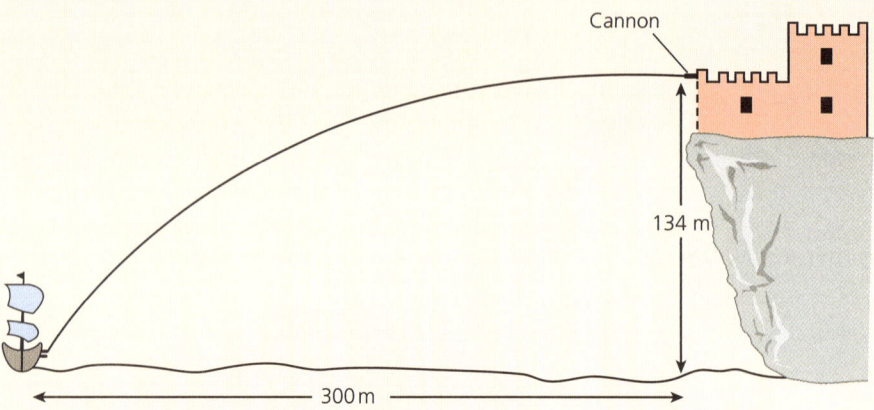

Figure 6 A cannon firing at an enemy ship

a i Assuming that air resistance is negligible, calculate the time of flight of the cannonball, from being fired horizontally to hitting the waterline of the ship.

2 marks

ii Use your answer to a i to calculate the initial horizontal muzzle velocity of the cannonball as it left the cannon.

2 marks

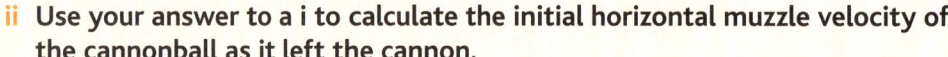

iii Use your answer to a i to calculate the vertical component of the velocity of the cannonball as it hit the ship.

2 marks

iv Use your answers to a ii and a iii, and a calculation or scale drawing, to determine the size and direction of the actual velocity of the cannonball as it hit the ship.

4 marks

b i The cannon had a mass of 1600 kg and fired cannonballs of mass 12 kg.
Calculate the loss of gravitational potential energy as the cannonball fell.　**2 marks**

....

....

....

ii Use your answer to a ii to calculate the recoil velocity of the cannon.　**3 marks**

....

....

....

....

....

....

iii Describe the energy changes that took place from the moment the
cannonball left the cannon until just before it hit the waterline of the ship.
Include the effects of air resistance in your answer.　**2 marks**

....

....

....

....

2 At the start of a race a sprinter settles into his blocks, as shown in Figure 7.　

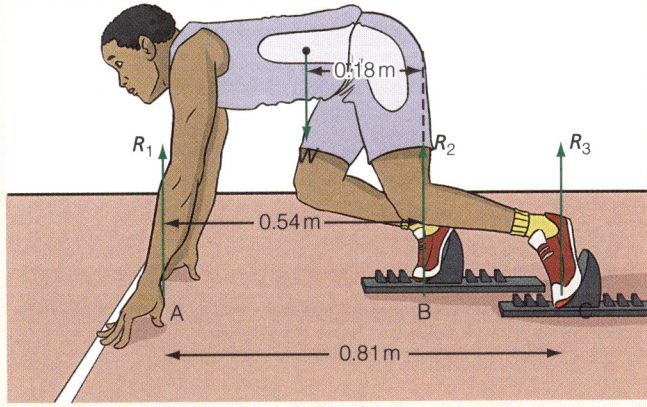

Figure 7 A sprinter in his blocks

Force R_1 is the resultant force on the sprinter's fingertips. The reaction force on his
forward foot, R_2, is 340 N and his weight W is 780 N. R_3 is the reaction force on his
back foot.

a i Calculate the moment of the sprinter's weight about his fingertips. Give an
appropriate unit for this quantity.　**2 marks**

....

....

....

ii Taking moments about his fingertips at A, calculate the force R_3 on his back foot.

3 marks

iii Calculate the force R_1 on his fingertips.

1 mark

b The sprinter starts running from his blocks and reaches a horizontal velocity of 11.5 m s^{-1} after a distance of 43 m.

i Calculate the average acceleration of the sprinter over this distance.

2 marks

ii Calculate the resultant force required to produce this acceleration.

2 marks

iii Use your answer to b ii to calculate the work done by the sprinter during this acceleration.

1 mark

iv The sprinter continues to exert the force calculated in b ii, but this is balanced by the forces of friction with the track and air resistance.
He continues to the 100 m finish line at his top horizontal velocity of 11.5 m s^{-1}.
Use this information to calculate his power from 43 m to the finish line.

1 mark

Topic 2 Materials

Bulk properties of solids

Density

The **density** of a material is fundamental to its use in the design and manufacture of objects. The density, ρ, of a material is defined as:

$$\text{density} = \frac{\text{mass}}{\text{volume}} \qquad \rho = \frac{m}{v}$$

The unit of density is kg m^{-3}.

Hooke's law

In 1678, the physicist Robert Hooke discovered that when a force, F, is applied to a spring, which extends by an amount Δl, these two quantities are proportional to each other, until the spring begins to permanently deform. This relationship is now known as **Hooke's law**:

$$F \propto \Delta l \quad \text{or} \quad F = k\Delta l$$

where k is a constant for that particular spring, called the **spring constant** or stiffness of the spring. Springs obey Hooke's law up to a point called the **limit of proportionality**, beyond which the extension is no longer proportional to the load force. The point at which the force produces a permanent stretch of the spring, and the spring does not return to its original length when the force is removed, is called the **elastic limit**. Up to this point, the spring shows **elastic deformation** and returns to its original length once the force is removed. Beyond the elastic limit, the spring shows **plastic deformation** and will have a permanent extension.

Tensile stress and strain

The **tensile stress**, σ, applied to an object is defined as the force, F, applied per unit cross-sectional area, A:

$$\text{tensile stress, } \sigma = \frac{F}{A}$$

Tensile stress is measured in pascals, Pa (or N m^{-2}).

The **tensile strain**, ε, applied to an object is defined as the extension of the material, Δl, per unit length, l:

$$\text{tensile stress, } \varepsilon = \frac{\Delta t}{l}$$

Tensile strain is a dimensionless quantity and is represented by a number only, it has no unit.

Elastic strain energy

When an object is subject to elastic deformation, work is done on the object by the tensile stress, and **elastic strain energy** may be stored in the object. For example, with a spring, the elastic strain energy is released when the stress is removed and the spring returns to its original shape. If the stress is too high, when the spring deforms, the elastic strain energy inside the spring will be higher than the internal forces holding the spring together and the spring will break. The stress at this point is called the **breaking stress**.

For an object obeying Hooke's law, the elastic strain energy stored in the object subject to a force, F, can be calculated using the equation:

$$\text{elastic strain energy stored} = \frac{1}{2} \times F \times \Delta l$$

where Δl is the extension produced by the force, F. This equation also calculates the area under a force–extension graph for an object that obeys Hooke's law, so:

$$\text{elastic strain energy stored} = \text{area under a force–extension graph}$$

Some materials obey Hooke's law, but then beyond the elastic limit of the material they break or **fracture**. Materials that behave like this are called **brittle** materials. Other materials, like rubber, behave differently under loading than they do when unloaded. As the elastic strain energy equals the area under the force–extension graph, the amount of energy stored in a rubber band under loading is bigger than the energy released when unloaded. The difference in area is the energy wasted as heat energy in the rubber band as the material is stretched and then released.

32 The springs in the seat of a car compress when a person sits on them. Figure 8 shows the force–extension graph for one of these springs.

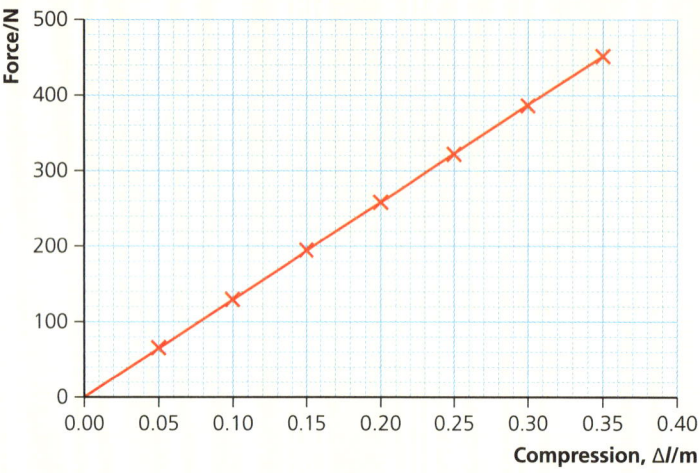

Figure 8

The spring constant and elastic strain energy stored in the spring when it is compressed by 0.35 m are given in Table 3. (AO2)

1 mark

Table 3

	Spring constant	Elastic strain energy
A	$1300\,N\,m^{-2}$	160 J
B	$7.8 \times 10^{-4}\,N\,m^{-2}$	79 J
C	$1300\,N\,m^{-2}$	79 J
D	$7.8 \times 10^{-4}\,N\,m^{-2}$	160 J

33 Figure 9 shows the force–extension graph of a thin bungee cord.

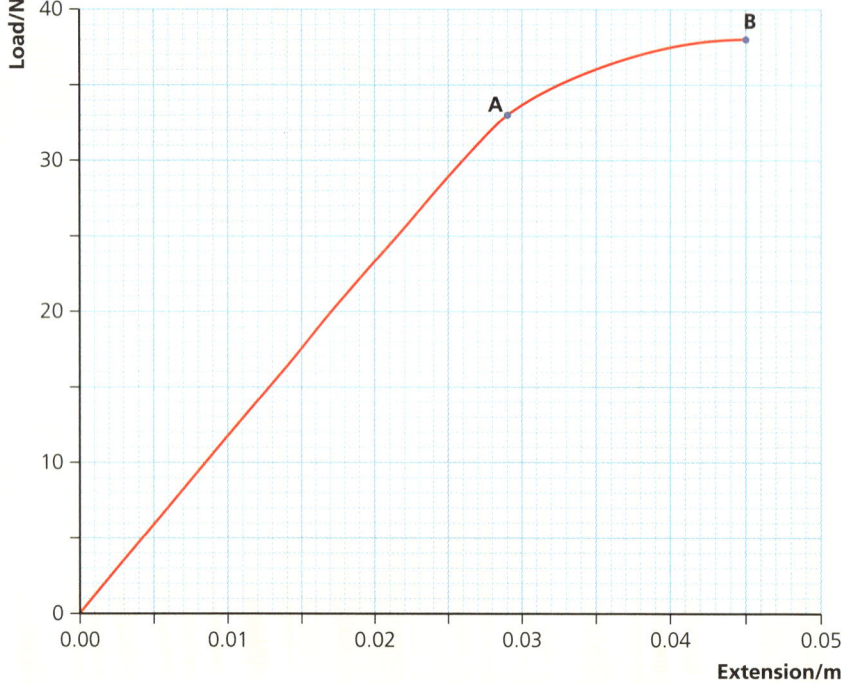

Figure 9

a   Calculate the spring constant, *k*, of the bungee cord. (AO2)   2 marks

b   Point A on the graph is called the elastic limit. Explain what is meant by the term 'elastic limit'. (AO1)   2 marks

....

....

....

....

c   Use the graph to determine the work done in stretching the bungee cord to point A. (AO2)   3 marks

d   Beyond point A, the bungee cord undergoes plastic deformation. Explain what is meant by the term 'plastic deformation'. (AO1)   1 mark

....

....

....

e   When the extension of the bungee cord reaches point B on the graph, the load force is reduced back to zero. On the graph, sketch how the extension of the bungee cord will vary with load force as the load is reduced to zero. (AO3)   2 marks

f   Without doing any extra calculations, compare the work done in unloading the load force from the bungee cord from point B to the force done in stretching the bungee cord to point B. (AO3)   1 mark

....

....

The Young modulus

The **Young modulus**, E, is a measure of the stiffness of a material and is defined as:

$$\text{Young modulus, } E = \frac{\text{tensile stress}}{\text{tensile strain}}$$

$$E = \frac{\sigma}{\varepsilon} = \frac{Fl}{A\Delta l}$$

where F is the force applied to the object, l is the length of the object, A is its cross-sectional area and Δl is the extension produced by the force F. The Young modulus is measured in pascals, Pa (or N m^{-2}).

Stress–strain graphs can be used to determine the Young modulus of materials as the gradient of the linear part of the graph equals the Young modulus. This can help us to represent the properties of the materials.

34 A stress–strain graph illustrating the mechanical behaviour of four different materials, A, B, C and D, is shown in Figure 10.

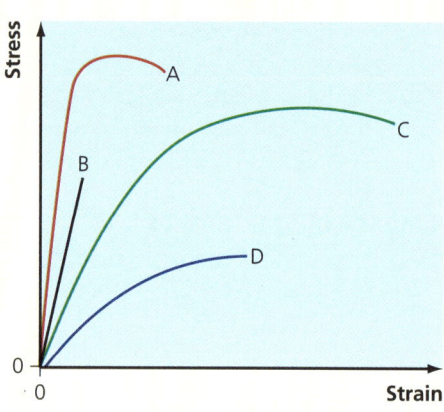

Figure 10

Which curve represents the most brittle material? (AO1)

1 mark

35 Table 4 shows the strain and corresponding stress values for a wire made from a steel alloy.

Table 4

Strain/10^{-3}	0	1.00	2.00	3.00	4.00	5.00	6.00	7.00	8.00	9.00	10.00
Stress/10^8 Pa	0	1.05	2.15	3.10	3.35	3.15	3.25	3.45	3.55	3.55	3.45

a Draw a graph of these data on the axes given in Figure 11 below. (AO2) 3 marks

Figure 11

b Use your graph to calculate a value for the Young modulus of the steel alloy. (AO2) 2 marks

c A steel rod 0.8 m long is required as part of the jack for a lorry. The largest tension in the rod will be 4.0 kN, and the rod cannot flex (extend) by more than 0.75 mm. If the Young modulus of the steel is 1.85×10^{11} Pa, calculate the minimum cross-sectional area of the rod. (AO2) 3 marks

Exam-style questions

1 **a** Describe an experiment to accurately measure the data required to determine the Young modulus of a copper wire. Include a diagram as part of your description. The quality of your written answer will be assessed as part of this question.

6 marks

....

....

....

....

....

....

....

....

....

....

....

....

b Figure 12 shows the stress–strain graph for a copper wire.

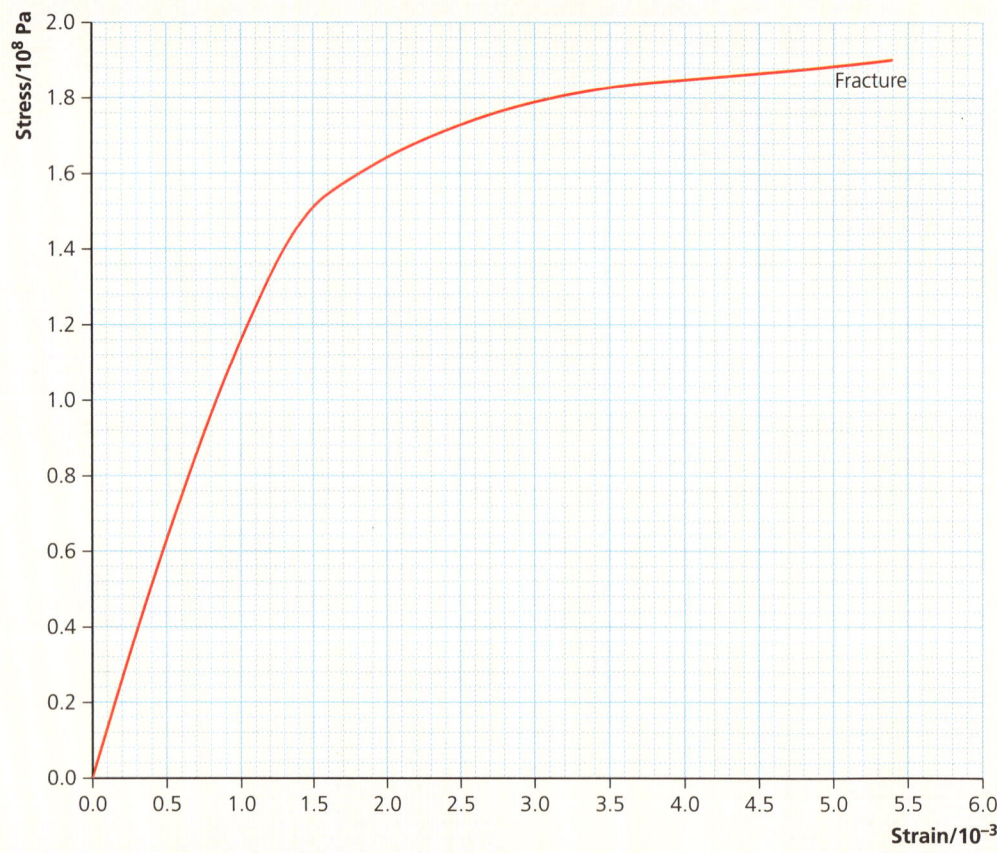

Figure 12

Use the graph to determine the breaking stress of the copper wire. `1 mark`

c On Figure 12, mark the point at which you consider plastic deformation to start. Label this point P. `1 mark`

d Use Figure 12 to calculate the Young modulus of copper. Include the correct unit as part of your answer. `3 marks`

e The area under a stress–strain graph represents the work done per unit volume to stretch the copper wire. Use Figure 12 to calculate the work done per unit volume of wire stretching the wire to a strain of 3.0×10^{-3}. `2 marks`

f Use your answer to e and the density of copper, $\rho = 8960\,\text{kg}\,\text{m}^{-3}$, to calculate the energy required to stretch a 0.030 kg sample of copper wire to a strain of 3.0×10^{-3}. `2 marks`

g A brittle plastic fractures at a tensile stress of 114 MPa and has a Young modulus less than that of copper. On Figure 12, sketch a possible stress–strain line for this plastic. `2 marks`

2 A tower crane is lifting building materials on a large building site, as shown in Figure 13. `14`

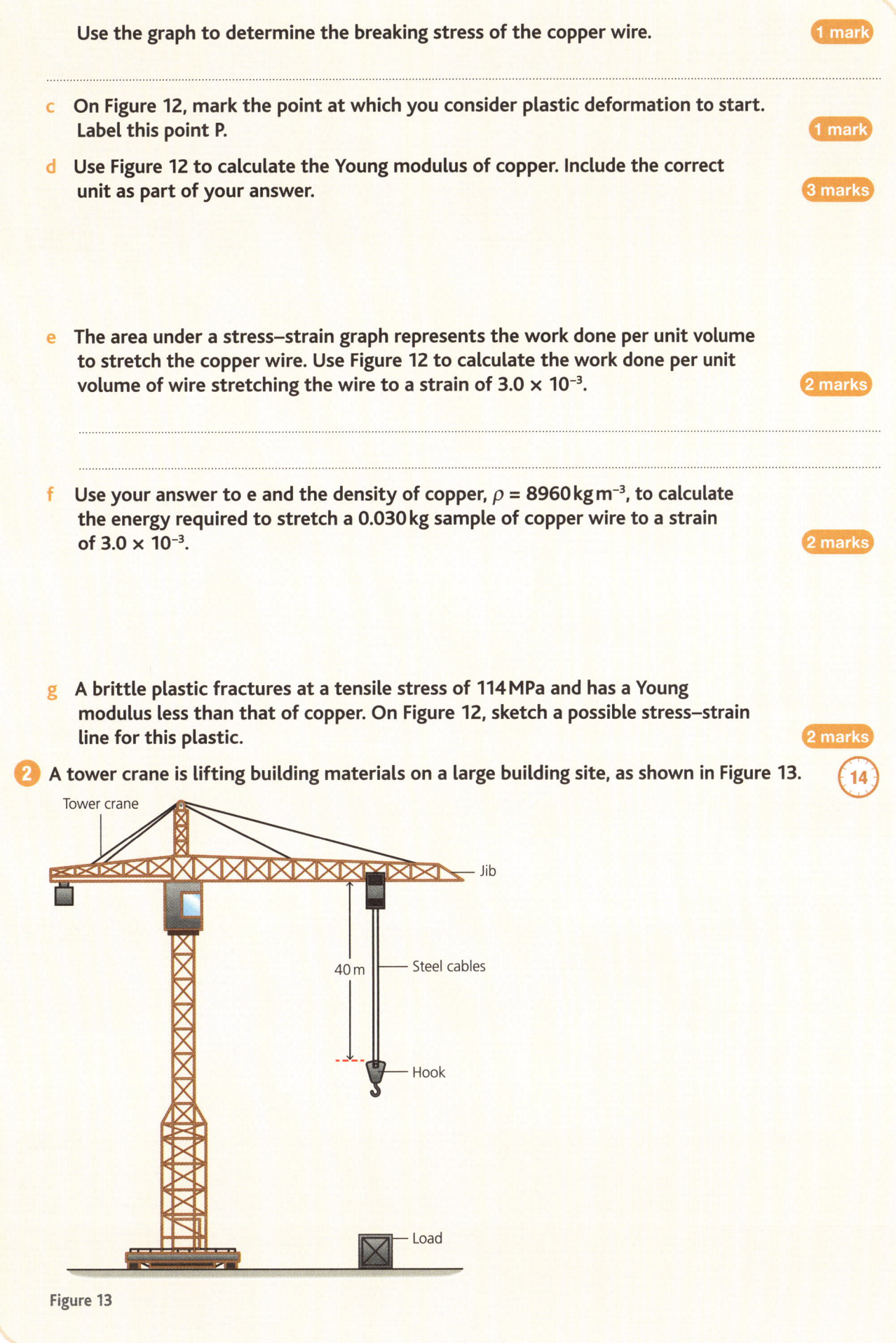

Figure 13

a Two identical steel cables, each 40 m long, attach a hook to the crane jib. Each cable has a mass of 5.1 kg per metre. Use these data to calculate the weight (in N) of one 40 m length of cable. **2 marks**

....

....

b The safe operating load of the crane is much less than 1.5×10^6 N, at which value the cables would break. If the cross-sectional area of each cable is 6.2×10^{-4} m², calculate the breaking stress of one cable. **2 marks**

c Under load each cable is subject to a stress of 400 MPa. Calculate the weight of the load. You may ignore the weight of the cable. **2 marks**

d The unstretched length of each cable is 40 m. Assuming that each cable obeys Hooke's law, and taking the Young modulus of steel to be 2.1×10^{11} Pa, calculate the extension of each cable under the load calculated in c. **3 marks**

e Use your answer to d to calculate the combined stiffness constant, k, for two cables. **2 marks**

f Calculate the total energy stored in the combined stretched cables. **2 marks**

Section 5 Electricity

Topic 1 Current electricity

Basics of electricity

Electric **current**, I, is the rate of flow of charge and is defined as:

$$I = \frac{\text{electric charge}}{\text{time to flow}} = \frac{\Delta Q}{\Delta t}$$

where ΔQ is the electric charge (measured in coulombs, C) that flows in a time interval, Δt (measured in s). Current is measured in amperes (amps), A.

Potential difference (pd), V, is the work done per unit charge and is defined as:

$$V = \frac{\text{work done}}{\text{charge}} = \frac{W}{Q}$$

where W is the electrical work done (measured in joules, J) and Q is the charge. Potential difference is measured in volts, V.

Resistance, R, is defined as:

$$R = \frac{\text{potential difference}}{\text{current}} = \frac{V}{I}$$

where R is the resistance measured in ohms, Ω.

---

36 Which of the following quantities could be measured in J s C^{-2}? (AO1) `1 mark`

   A   current   C   potential difference

   B   work done   D   resistance

37 A circuit for a simple LED torch consists of a 3.0 V battery, of negligible internal resistance, connected in series with an LED and a 1.2 kΩ fixed resistor. When operating normally, the LED has a potential difference of 1.6 V across it.

   a   Calculate the potential difference across the 1.2 kΩ resistor. (AO2) `1 mark`

   b   Use your answer to a to calculate the current in the circuit. (AO2) `2 marks`

   c   Calculate the resistance of the LED when the torch is on. (AO2) `2 marks`

   d   Calculate the total charge that flows through the LED in 5 minutes. (AO2) `2 marks`

   e   Calculate the total work done by the charge flowing through the LED in 5 minutes. (AO2) `2 marks`

29

Current–voltage characteristics

The best way to analyse the behaviour of an electrical component is to plot a current–voltage (or potential difference) graph, which shows the **electrical characteristic** (or current–voltage characteristic) of the component. The electrical characteristic of a filament lamp is shown in Figure 14.

As shown in Figure 14, electrical characteristics can be drawn with the variables on either axis.

Components such as fixed resistors (and metal wires), at a constant temperature, have linear current–voltage characteristics, where $I \propto V$. These are called **ohmic conductors**. The current–voltage characteristic of a light-emitting diode (LED) is non-linear and, as such, this component is called a **non-ohmic conductor**.

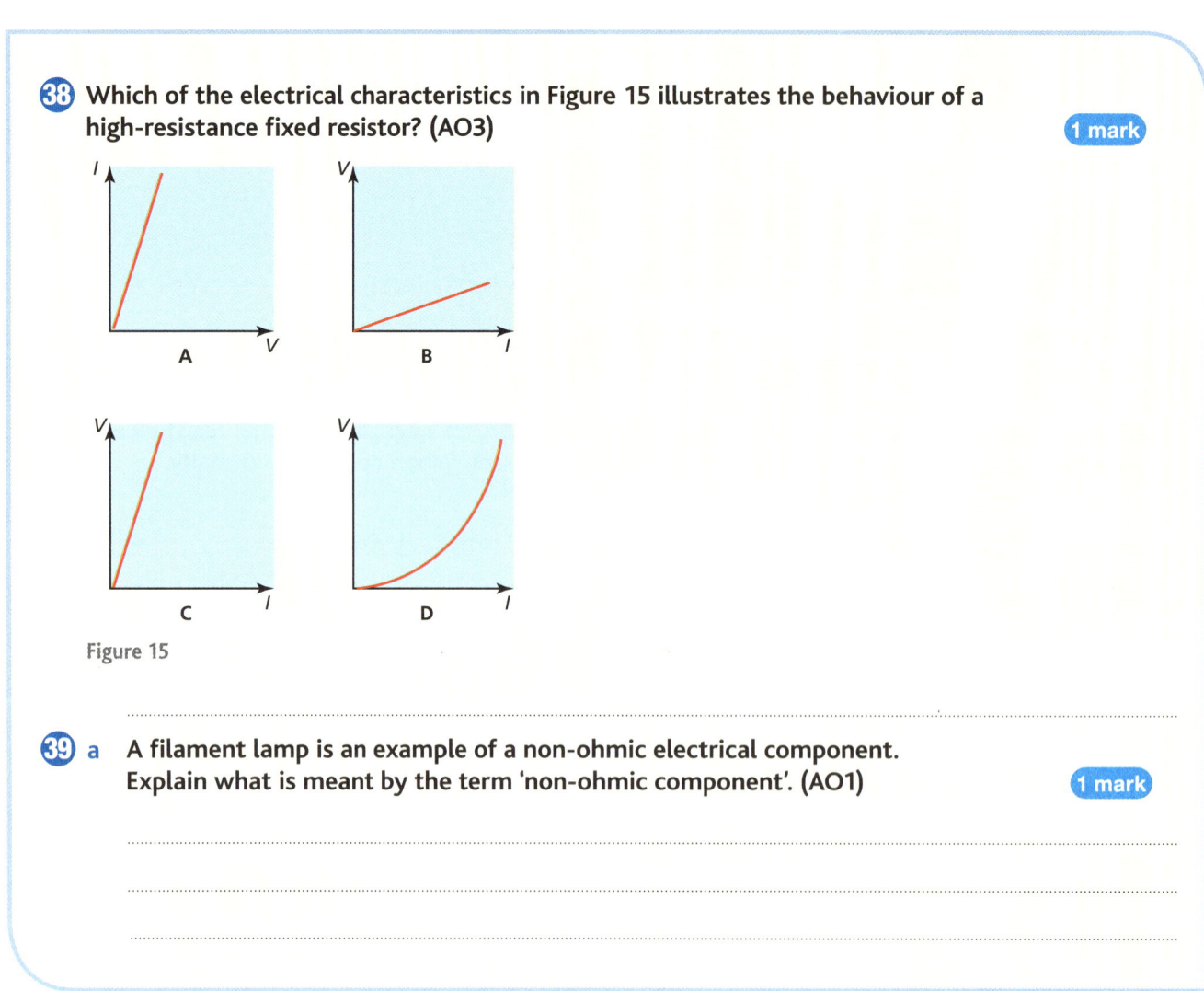

Figure 14 The electrical characteristic of a filament lamp

38 Which of the electrical characteristics in Figure 15 illustrates the behaviour of a high-resistance fixed resistor? (AO3) 1 mark

Figure 15

39 a A filament lamp is an example of a non-ohmic electrical component. Explain what is meant by the term 'non-ohmic component'. (AO1) 1 mark b   Sketch the current–voltage characteristic of a filament lamp on a set of axes with *I* on the *y*-axis and *V* on the *x*-axis. (AO1)   2 marks

c   Using the current–voltage characteristic that you have drawn in b, state how the resistance of a filament lamp changes with applied pd. (AO3)   1 mark

....

....

d   The current–voltage characteristic of a fixed resistor is shown in Figure 16.

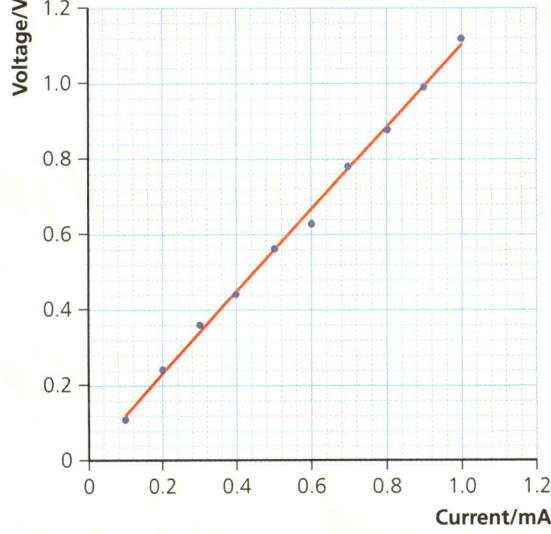

Figure 16

Use the electrical characteristic in Figure 16 to calculate the resistance of the fixed resistor. (AO2)   3 marks

....

....

....

....

....

....

Resistivity

Increasing the length of a conductor increases its resistance, and increasing its width decreases its resistance. The **resistivity** of a material is the intrinsic resistance of a material, independent of its dimensions, and is defined by:

$$\rho = \frac{RA}{l}$$

where R is the resistance of the conductor (in Ω), A is its cross-sectional area (in m²) and l is its length (in m). Resistivity is measured in ohm metres, Ωm.

As well as dimensions, the resistance of a metal conductor also depends upon temperature. Increasing temperature increases the resistance of a metal wire. This is due to the increased rate of collisions between the conducting electrons and the ions (or particles)

within the metal. Thermistors are components with a resistance that is temperature dependent. Negative temperature coefficient (or ntc) thermistors are components made from semiconductors, and their resistance **decreases** with increasing temperature. At higher temperatures, more conducting electrons are released from the semiconducting, material, allowing for greater conduction and therefore lower resistance.

Superconductors are a class of materials whose resistance, and therefore their resistivity, drops to zero below a certain critical temperature, T_c. The critical temperature depends upon the material that the superconductor is made from, and values range from just above absolute zero (–273.15 °C) for materials such as tungsten, to about 130 K (–143 °C) for high temperature superconducting ceramics.

---

40 The resistance of a superconductor when it is superconducting depends upon its: (AO1) `1 mark`

- A   length
- B   diameter
- C   cross-sectional area
- D   critical temperature

---

41 A nichrome wire has a diameter of 0.15 mm and length of 85 cm. An ohmmeter measures its resistance to be 53 Ω.

a   Calculate the cross-sectional area of the wire in m². (AO2)   `1 mark`

b   Calculate the resistivity of the wire. (AO2)   `2 marks`

c   The wire is replaced by one that is double the length with half the cross-sectional area. What is the resistance of this wire? (AO2)   `2 marks`

d   The original wire is replaced by a similar wire from the same reel, but the wire is 1.54 m long. What is the resistance of this wire? (AO2)   `2 marks`

Circuits

Resistors

When resistors are combined in a **series** circuit, the total resistance of the circuit, R_T, is the sum of the individual resistors in the circuit:

$R_T = R_1 + R_2 + R_3 + \dots$ (for more resistors)

When resistors are combined in a **parallel** circuit, the total resistance, R_T, is determined using:

$$\frac{1}{R_T} = \frac{1}{R_1} + \frac{1}{R_2} + \frac{1}{R_3} + \dots$$

Energy and power

The energy, E, transferred by electrical components such as heaters, lamps and motors is given by:

$E = VIt$

The electrical power dissipated by a component, P, is therefore given by:

$$P = \frac{E}{t} = \frac{VIt}{t} = VI = I^2R = \frac{V^2}{R}$$

Conservation of energy and charge

In **parallel** circuits with junctions, current is conserved at each junction, which means that:

total current flowing into a junction
= total current flowing out of a junction

In **series** circuits (or complete loops in a parallel circuit) containing a power supply, energy is conserved. In other words, for a constant current, because the energy transferred by a component is proportional to the voltage (or pd) across it:

total voltage put into a circuit
= total voltage taken out of a circuit

The voltage supplied to a circuit (when no current is flowing) by a power supply, such as a battery, is called the **electromotive force** (emf), ε, where:

$$\varepsilon = \frac{E}{Q}$$

This is the energy per unit charge. The voltage removed from a circuit by a component, such as a lamp or a resistor, is called the **potential difference**, V, thus:

sum of pds supplied around any closed circuit loop
= sum of pds used around any closed circuit loop

---

 42 **A battery of emf 12 V, with negligible internal resistance, is connected to the resistor network shown in Figure 17. (AO2)** `1 mark`

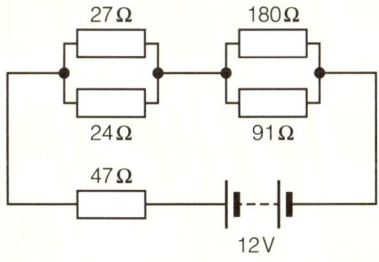

Figure 17

The current flowing through the 91 Ω resistor is:

A 0.07 A

B 0.09 A

C 0.03 A

D 0.14 A

43 Figure 18 shows three resistors connected to a battery of negligible internal resistance.

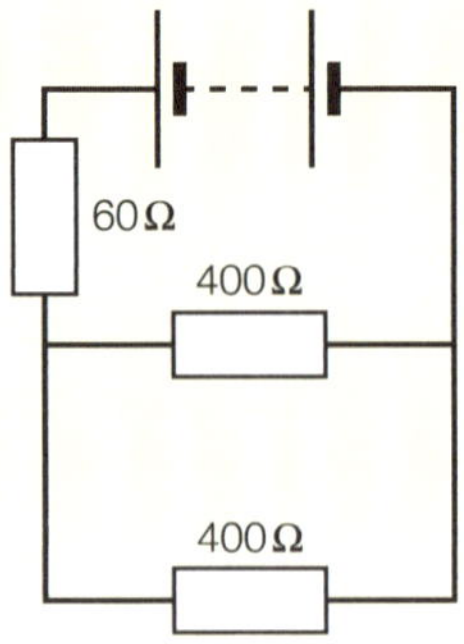

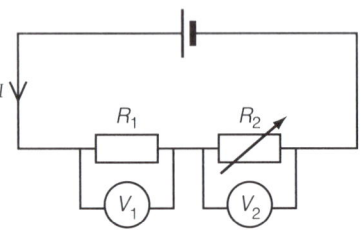

Figure 18

a   Calculate the total resistance of the circuit. (AO2)   [2 marks]

b   The power dissipated by each of the 400 Ω resistors is 1.5 W. Calculate the pd across the two 400 Ω resistors. (AO2)   [2 marks]

c   Calculate the current through the 60 Ω resistor. (AO2)   [2 marks]

d   Calculate the emf of the battery. (AO2)   [2 marks]

....

....

Potential divider

A **potential divider** is a circuit that can be used to control the potential differences within a circuit or be used to measure the changing potential difference produced by a sensor, such as a thermistor or light dependent resistor (LDR). The circuit shown in Figure 19 shows a simple potential divider circuit involving a fixed resistor and a variable resistor, which can be used to vary and control the potential differences within the circuit.

The components in a potential divider circuit, such as the one shown in Figure 19, are all connected in series, therefore:

$$\varepsilon = V_1 + V_2$$

However, as the current is the same throughout the circuit, V_1 and V_2 depend upon the values of R_1 and R_2. Changing the value of either resistance will affect the values of V_1 and V_2. Figure 19 is set up as a voltage-controlling circuit, but the variable resistor could be replaced by a thermistor or an LDR, and the circuit could then be used to measure changes in temperature or light intensity.

Figure 19 A circuit diagram for a potential divider

44 A resistor and a thermistor are connected in series with a 12 V battery of negligible internal resistance, and are used as part of a temperature-sensing circuit. At 150°C, the resistance of the thermistor is 9.0 kΩ and the resistance of the resistor is 65 kΩ. What is the pd across the thermistor at this temperature? (AO2) **1 mark**

    A   10 V

    B   11 V

    C   1.1 V

    D   1.5 V

45 A room temperature sensor circuit consists of a 6.0 V battery of negligible internal resistance connected in series to a low-resistance ammeter and a 5.0 kΩ resistor, and is then connected to an ntc (negative temperature coefficient) thermistor and a 10 kΩ resistor connected in parallel to each other, before returning to the battery, as shown in Figure 20.

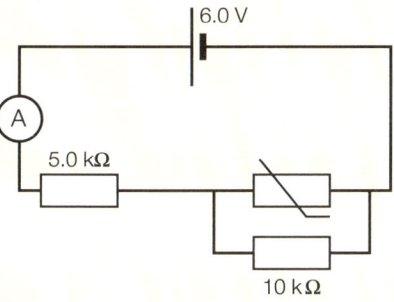

Figure 20

At its normal working room temperature (20°C), the reading on the ammeter is 857 µA.

a Calculate the pd across the 5.0 kΩ resistor. (AO2) **2 marks**

b Calculate the pd across the 10 kΩ resistor. (AO2) **2 marks**

c Calculate the combined resistance of the thermistor and the 10 kΩ resistor parallel combination. (AO2) **2 marks**

d Calculate the resistance of the thermistor. (AO2) **3 marks**

e The temperature inside the room starts to drop. State and explain the effect that this has on the ammeter reading. (AO3) **3 marks**

Electromotive force and internal resistance

All real sources of electrical energy, such as cells, power supplies and generators, have an **internal resistance**. This internal resistance produces a drop in potential difference inside the power supply that cannot be measured directly because of the energy loss (mainly as heat) inside the power source. The small pd produced by the internal resistance, r, means that the terminal pd, V, supplied to the external load, R, becomes less than the original emf, E, of the power source. This means that:

$$\varepsilon = V + Ir \qquad \text{or} \qquad \varepsilon = IR + Ir$$

46 A battery has an emf of 12 V and an internal resistance of 2.0 Ω, and is connected to two 30 Ω resistors in series. The pd across the internal resistance is: (AO2) `1 mark`

A   0.34 V   B   0.36 V   C   0.39 V   D   0.40 V

47 A variable resistor, Y, was connected in series to a battery of emf, ε, and internal resistance, r. The power dissipated through Y for different values of resistance of Y was measured, and is shown in Figure 21 for resistance values between 0.5 Ω and 6.5 Ω.

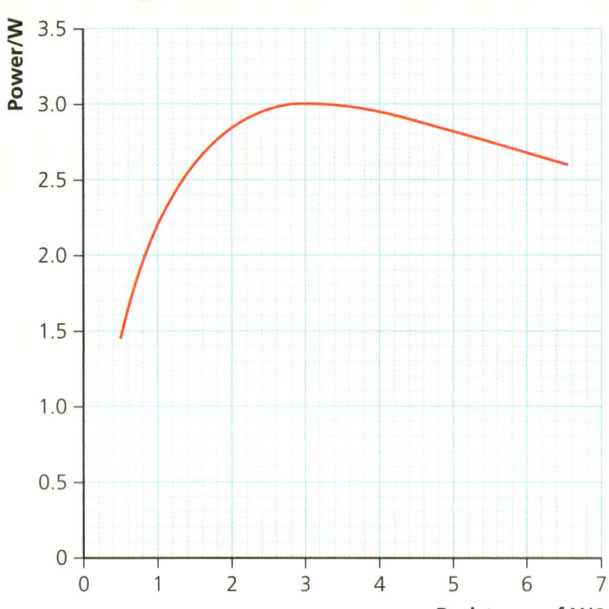

a Use the shape of the graph to describe how the power dissipated in Y varies between these two values of resistance. (AO3) `2 marks`

Figure 21

b The emf of the battery, ε, is 6.0 V and the resistance of Y is set to 0.80 Ω. Use Figure 21 to determine the current flowing through the battery. (AO2) `3 marks`

c Use your value from b to calculate the pd across Y. (AO2) `2 marks`

d Hence calculate the internal resistance, r, of the battery. (AO2) `2 marks`

Exam-style questions

1 A battery of emf 12 V and 1.5 Ω internal resistance is connected in parallel to a 2.0 Ω resistor and an unknown resistor, R. The external parallel circuit draws a current of 4.2 A from the battery, as shown in Figure 22.

 13

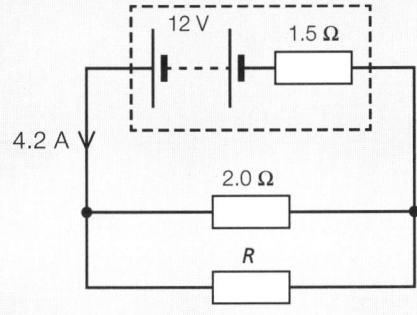

Figure 22

a Calculate the pd across the internal resistance. **1 mark**

....

....

b Use your answer to a to determine the pd across the 2.0 Ω resistor. **1 mark**

....

....

c Hence calculate the current in the 2.0 Ω resistor. **1 mark**

d Calculate the current flowing through resistor R. **1 mark**

e Show that the resistance of R is 4.2 Ω. **1 mark**

f Determine the total resistance of the circuit. **2 marks**

g The chemical energy stored in the battery is gradually transformed into electrical energy within the circuit, which is dissipated as heat by the internal resistance and the two external resistors. Use data from a to f to calculate the rate of energy dissipation in each of the three resistors. **3 marks**

....

....

....

....

h   Use your answer to g to show that energy is conserved in this circuit. `3 marks`

2   An experiment is carried out to investigate how the resistance of a metal wire changes between the temperatures of 0°C and 100°C. `25`

   a   Draw a labelled diagram of the experimental set-up that would enable this experiment to be carried out. `3 marks`

   b   Describe the experimental procedure needed to take accurate and reliable measurements of the resistance of the wire between these two temperatures. You will be assessed on the quality of your written communication. `6 marks`

   c   The critical temperature of tin is −269.43°C (3.72 K). Explain what is meant by the term 'critical temperature'. `2 marks`

d X and Y are two different lamps. X is rated as 12 V, 36 W. Y is rated as 4.5 V, 2.0 W. The two lamps are connected to resistors R_1 and R_2, as shown in Figure 23.

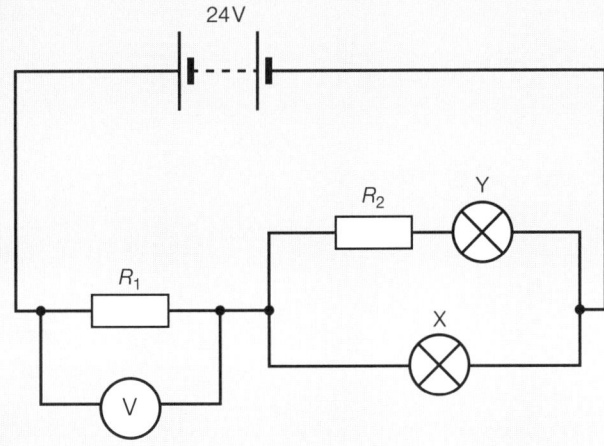

Figure 23

The 24 V battery has negligible internal resistance, and the values of R_1 and R_2 are chosen so that lamps X and Y operate at their stated working pds. Calculate the current in each lamp when it is operated at its correct working pd.

e Use the data and the circuit diagram to complete Table 5.   6 marks

Table 5

Resistor	Pd across resistor/V	Current flowing through resistor/A	Resistance of resistor/Ω
R_1			
R_2			

f The filament wire inside lamp X breaks. Lamp X goes out. Lamp Y gets brighter and the reading on the voltmeter decreases. Explain why the reading on the voltmeter decreases. **2 marks**

....

....

....

g Explain why lamp Y gets brighter. **2 marks**

....

....

....

....

Philip Allan, an imprint of Hodder Education, an Hachette UK company, Blenheim Court, George Street, Banbury, Oxfordshire OX16 5BH

Orders
Bookpoint Ltd, 130 Milton Park, Abingdon, Oxfordshire OX14 4SB
tel: 01235 827827
fax: 01235 400401
e-mail: education@bookpoint.co.uk
Lines are open 9.00 a.m.–5.00 p.m., Monday to Saturday, with a 24-hour message answering service. You can also order through www.hoddereducation.co.uk

© Jeremy Pollard 2015
ISBN 978-1-4718-4504-8
First printed 2015
Impression number   5   4   3
Year   2020   2019   2018

Cover photo reproduced by permission of Fotolia
Typeset by Aptara, Inc.

Printed in Dubai

Hachette UK's policy is to use papers that are natural, renewable and recyclable products and made from wood grown in sustainable forests. The logging and manufacturing processes are expected to conform to the environmental regulations of the country of origin.

ISBN 978-1-4718-4504-8